U.S. ENVIRONMENTAL PROTECTION AGENCY

OFFICE OF INSPECTOR GENERAL

EPA Did Not Conduct Thorough Biennial User Fee Reviews

Report No. 14-P-0129   March 4, 2014

Report Contributors: Paul Curtis
                                   Arthur Budelier
                                   Sheree James
                                   Denise Patten
                                   Wendy Swan

Abbreviations

CFO	Chief Financial Officer
CFO Act	Chief Financial Officers Act of 1990
EPA	U.S. Environmental Protection Agency
FY	Fiscal year
GAO	U.S. Government Accountability Office
IOAA	Independent Offices Appropriation Act
NPDES	National Pollutant Discharge Elimination System
OCFO	Office of the Chief Financial Officer
OFM	Office of Financial Management
OIG	Office of Inspector General
OMB	Office of Management and Budget
ORD	Office of Research and Development
SFFAS	Statement of Federal Financial Accounting Standards

Hotline

To report fraud, waste or abuse, contact
us through one of the following methods:

email: OIG_Hotline@epa.gov
phone: 1-888-546-8740
fax: 1-202-566-2599
online: http://www.epa.gov/oig/hotline.htm

write: EPA Inspector General Hotline
          1200 Pennsylvania Avenue, NW
          Mailcode 2431T
          Washington, DC 20460

Suggestions for Audits or Evaluations

To make suggestions for audits or evaluations,
contact us through one of the following methods:

email: OIG_WEBCOMMENTS@epa.gov
phone: 1-202-566-2391
fax: 1-202-566-2599
online: http://www.epa.gov/oig/contact.html#Full_Info

write: EPA Inspector General Hotline
          1200 Pennsylvania Avenue, NW
          Mailcode 2431T
          Washington, DC 20460

At a Glance

EPA Did Not Conduct Thorough Biennial User Fee Reviews

What We Found

The EPA did not conduct thorough biennial user fee reviews for fiscal years 2008–2009 and 2010–2011, and did not review all agency programs to determine whether they should assess fees for government services they provide. The EPA did not fully comply with the requirements to:

> The EPA may not have recovered all program costs and collected millions of dollars that could have been available to reduce the federal budget deficit.

- Conduct cost reviews to determine the full cost of providing a service.
- Report biennial review results to OMB.
- Request user fee exceptions by letter to the OMB Director.
- Review all programs for fee potential.

The EPA's OCFO did not fully oversee the biennial reviews or provide internal review guidance, and the EPA's program offices were not fully aware of biennial review requirements. Consequently, the EPA may not have recovered millions of dollars in program costs and collected funds that could have been available to reduce the federal budget deficit. We identified an EPA program—the Office of Water's National Pollutant Discharge Elimination System (NPDES) program—with the potential to charge fees up to $8.9 million per year to recover its costs of providing a service.

The EPA began improving its biennial review process with the fiscal years 2012–2013 review by issuing a biennial user fee review guide, training user fee program personnel on biennial reviews, and increasing headquarters oversight of reviews.

Recommendations and Planned Agency Corrective Actions

We recommend that the CFO discuss biennial user fee results in the Agency Financial Report, coordinate requests for an exception to charging fees, and request fee exception programs to provide complete information about program fees and costs and help determine whether fees should be assessed. We also recommend that the Office of Water conduct an analysis to determine the EPA's full cost of issuing NPDES permits and determine whether it should charge fees for the permits. We had also recommended that the Office of Water propose a regulation to allow the EPA to charge NPDES permit fees, as appropriate.

The agency concurred and provided acceptable corrective actions and milestone completion dates for all recommendations except one—to propose a regulation to allow the EPA to charge NPDES permit fees. We revised our recommendation by removing the proposal for a regulation to charge NPDES fees and adding the option for requesting an exception to fees, to which the Office of Water agreed.

UNITED STATES ENVIRONMENTAL PROTECTION AGENCY
WASHINGTON, D.C. 20460

March 4, 2014

MEMORANDUM

SUBJECT: EPA Did Not Conduct Thorough Biennial User Fee Reviews
Report No. 14-P-0129

FROM: Arthur A. Elkins Jr.

TO: Maryann Froehlich, Acting Chief Financial Officer
Office of the Chief Financial Officer

Nancy K. Stoner, Acting Assistant Administrator
Office f Water

This is our report on the subject audit conducted by the Office of Inspector General (OIG) of the U.S. Environmental Protection Agency (EPA). This report contains findings that describe the problems the OIG has identified and corrective actions the OIG recommends. The offices we identified with primary jurisdiction over the audit issues and the responsibility for taking corrective action are the Office of the Chief Financial Officer's Office of Financial Management and the Office of Water's Office of Wastewater Management. This report represents the opinion of the OIG and does not necessarily represent the final EPA position. Final determinations on matters in this report will be made by EPA managers in accordance with established audit resolution procedures.

Action Required

The agency agreed with recommendations 1 through 4 and completed the corrective actions. These recommendations are closed and no further action is required. The agency stated that it agrees with our revised recommendation 5 and took the corrective action to request the Office of Management and Budget (OMB) for an exception to charging National Pollutant Discharge Elimination System program fees. Therefore, you are not required to provide a written response for recommendation 5. However, it will remain open until the OMB approves your request. Should you choose to provide a final response, we will post your response on the OIG's public website, along with our memorandum commenting on your response. You should provide your response as an Adobe PDF file that complies with the accessibility requirements of Section 508 of the Rehabilitation Act of 1973, as amended. We will post this report to our website at http://www.epa.gov/oig.

If you or your staff have any questions regarding this report, please contact Richard Eyermann, acting Assistant Inspector General for Audit, at (202) 566-0565 or eyermann.richard@epa.gov; or Paul Curtis, Product Line Director for Financial Statement Audits, at (202) 566-2523 or curtis.paul@epa.gov.

Table of Contents

Chapters

Appendices

Chapter 1
Introduction

Purpose

We performed this audit to evaluate the U.S. Environmental Protection Agency's (EPA's) biennial user fee reviews. With the Administration's current focus on reducing the federal budget deficit, we wanted to determine whether the EPA was conducting effective reviews of its user fee programs. The objectives of our audit were to determine whether the EPA:

- Conducts biennial reviews of the EPA's user fees and royalties programs.
- Reviews all agency programs to determine whether fees should be assessed for government services they provide.

Background

In recent Office of Inspector General (OIG) reviews of two EPA user fee programs, we found that the programs were not conducting biennial cost reviews[1] as required by the Chief Financial Officers Act of 1990 (CFO Act) and as directed by Office of Management and Budget (OMB) Circular A-25. Without a cost study, the EPA programs did not have the cost data necessary to determine whether they should adjust their fees. Based on the findings in those reports, we conducted this additional work to determine the extent of biennial user fee reviews for all EPA fee programs.

For budget purposes, the U.S. Government Accountability Office (GAO) defines user fees as fees assessed on users for goods or services provided by the federal government.[2] User financing, in the form of user fees, is one approach to financing federal programs or activities. User fees assign part or all of the costs of these programs and activities—the cost of providing a benefit that is above and beyond what is normally available to the general public—to readily identifiable users of those programs and activities. Because user fees represent a charge for a service or benefit received from a government program, payers may expect a tight link between their payments and the cost of providing services.

---

[1] EPA OIG reports, *EPA Should Update Its Fees Rule to Recover More Motor Vehicle and Engine Compliance Program Costs*, Report No. 11-P-0701, September 23, 2011; and *EPA Is Not Recovering All Its Costs of the Lead-Based Paint Fees Program*, Report No. 13-P-0163, February 20, 2013.

[2] See GAO, *A Glossary of Terms Used in the Federal Budget Process*, GAO-05-734SP (Washington, D.C.: September 2005).

Statutory Authorities

The Independent Offices Appropriation Act (IOAA) of 1952 authorizes federal agencies to charge fees for the services they provide. The IOAA requires that each charge be fair and based on the costs to the government, the value of the service to the recipient, the public policy or interest served, and other relevant facts. The IOAA states that each service provided by a federal agency should be self-sustaining to the extent possible. In many instances, Congress has provided specific statutory authority to federal agencies to assess user fees.

The CFO Act requires the Chief Financial Officer (CFO) to review, on a biennial basis, the fees, royalties, rents and other charges imposed by the agency for services and things of value it provides. The CFO shall make recommendations on revising those charges to reflect costs incurred by the agency in providing those services and things of value.

Federal Policy, Standards and Guidance

OMB Circular A-25, *User Charges*, dated July 8, 1993, implements Title V of the IOAA. It establishes federal policy regarding charges for government goods and services that convey special benefits to recipients beyond those accruing to the general public. It establishes that user charges should be set at a level sufficient to recover the full cost of providing the service, resource or good. It requires the agency to review the user charges for agency programs biennially, to include (1) assurance that existing charges are adjusted to reflect unanticipated changes in costs or market values, and (2) a review of all other agency programs to determine whether fees should be assessed for government services or the user of government goods or services. Agencies will generally implement user charges through the promulgation of regulations. When there are statutory prohibitions or limitations on charges, agencies should propose legislation to permit charges to be established. Agencies should discuss the results of the biennial review of user fees and any resultant proposals in the Agency Financial Report.

The Statement of Federal Financial Accounting Standards (SFFAS) No. 4, *Managerial Cost Accounting Standards and Concepts*, dated July 31, 1995, provides that full cost should be considered as a primary basis for setting fees for government goods and services. The full cost of an output is the total amount of resources used to produce the output, including direct and indirect costs. Indirect costs are costs that are jointly or commonly used to produce two or more types of outputs but are not specifically identifiable with any of the outputs. Typical examples of indirect costs include general and administrative services; general research and technical support; security; rent; employee health and recreation facilities; and operating and maintenance costs for buildings, equipment and utilities.

The GAO's *Standards for Internal Controls in the Federal Government* requires that internal controls provide reasonable assurance of reliable financial reporting. GAO's report, *Federal User Fees – A Design Guide*, GAO-08-386SP, dated May 2008, states that agencies must substantively review and report on their fees on a regular basis. This is to ensure that Congress, stakeholders and agencies have complete information about changing program costs and whether authorized activities align with program activities. Transparent processes for reviewing and updating fees help assure payers and other stakeholders that fees are set fairly and accurately and are spent on the programs and activities Congress intended.

Responsible Offices

The offices we identified with primary jurisdiction over the audit issues and the responsibility for taking corrective action on our recommendations are the Office of Financial Management within the Office of the Chief Financial Officer (OCFO) and the Office of Wastewater Management within the Office of Water.

Scope and Methodology

We conducted this audit in accordance with generally accepted government auditing standards. Those standards require that we plan and perform the audit to obtain sufficient, appropriate evidence to provide a reasonable basis for our findings and conclusions based on our audit objectives. We believe that the evidence obtained provides a reasonable basis for our findings and conclusions based on our audit objectives. We conducted our audit from February through August 2013. Appendix A contains details on our scope and methodology.

Chapter 2
EPA Should Conduct More Thorough Biennial User Fee Reviews

The EPA did not conduct thorough biennial user fee reviews for fiscal years (FYs) 2008–2009 and 2010–2011. Although the EPA prepared a biennial user fee review report, the EPA did not fully comply with the requirements to:

- Conduct cost reviews to determine the full cost of providing a service.
- Report biennial review results to OMB.
- Request user fee exceptions by letter to the OMB Director.
- Review all programs for fee potential.

The CFO Act and OMB Circular A-25 require the agency to perform biennial cost reviews of agency user fee program charges and make recommendations on revising those charges to reflect agency costs incurred. The EPA's OCFO did not fully oversee the biennial reviews or provide internal review guidance, and EPA program offices were not fully aware of the biennial review requirements. By not performing fully compliant biennial user fee reviews, the EPA may not have recovered all related program costs and collected funds that otherwise could have been available to reduce the federal budget deficit. We identified an EPA program—the Office of Water's National Pollutant Discharge Elimination System (NPDES) program—with the potential to charge fees up to $8.9 million per year to recover its costs of providing a service. The EPA began improving its biennial review process with the FYs 2012–2013 review by issuing a biennial user fee review guide, training user fee program personnel to perform biennial reviews, and increasing headquarters oversight of the reviews.

EPA Programs Generally Did Not Conduct Cost Reviews

The EPA's programs that charge user fees, and programs with exceptions from charging user fees, generally did not conduct cost reviews to determine the full cost of providing a service. The programs need cost reviews to determine their actual costs and decide whether they need to adjust fees to reflect changes in costs or propose new charges, as required by OMB Circular A-25.

OCFO's Office of Financial Management (OFM) had oversight responsibility for the EPA's biennial user fee reviews but did not provide sufficient oversight. OFM requested the EPA's programs to provide updated fee descriptions and indicate whether fees (1) recover the full cost of the service provided, and (2) were revised or adjusted to reflect changes in the cost or value of the service. Since the user fee programs generally did not conduct cost reviews, the programs could not indicate whether they recovered the full cost of the service provided or revised fees to reflect changes in the costs. OFM did not fully evaluate the information it received or follow up with the programs to obtain all the cost information it needed. OFM's biennial user fee report to OMB did not indicate whether fees recovered the full cost of the service provided, and for some types of fees whether the fees were revised or adjusted.

Some fee programs—including pesticides maintenance, pesticides registration, Clean Air Act Part 71 Operating Permits and Freedom of Information Act—performed a cost analysis to comply with reporting requirements other than the biennial review requirement. Some of these cost analyses did not include all indirect costs to obtain the full cost of their services, as directed by SFFAS No. 4 and OMB Circular A-25.

EPA's Office of Research and Development (ORD) did not conduct biennial cost reviews for its programs that collected licensing fees and royalties. However, ORD may not be subject to the OMB Circular A-25 requirement to biennially review program costs because ORD does not base its licensing fees and royalties on total program costs. OMB Circular A-25 provides guidance regarding the assessment of user charges under statutes other than the IOAA, but only to the extent permitted by law and not inconsistent with the statute. ORD programs generally establish their fees by individual agreement, based on specific statutory authority instead of full cost recovery guidance in SFFAS No. 4 and OMB Circular A-25. For example:

- The Stevenson–Wydler Technology Innovation Act of 1980 promotes the use of federally funded technology developments by state and local governments and the private sector. ORD collects licensing fees and royalties from negotiated license agreements on patented technologies that the EPA owns. ORD receives an up-front negotiated licensing fee and annual royalties based on a percentage or amount of the licensee's profit from sale or use of the technology.

- The Environmental Research, Development, and Demonstration Authorization Act of 1980 (42 U.S. Code § 4370), *Reimbursement for Use of Facilities*, authorizes the Administrator to allow appropriate use of EPA research and test facilities by outside groups or individuals and allows the agency to receive reimbursement for costs incurred or waive reimbursement for nonprofit private or public entities when the Administrator finds this to be in the public interest. ORD uses outside user's agreements to charge users for use of its research and test facilities.

- The Federal Technology Transfer Act of 1986 authorizes government laboratories to enter into cooperative research and development agreements with universities and the private sector for technological transfer for commercial purposes. These agreements are for joint research projects where both parties receive benefits. When a fee is involved, it usually reimburses the EPA only for the cost of supplies, travel or contractor costs.

From the EPA's list of programs with user fee exceptions, we identified the NPDES program as a potential fee program. The NPDES program has the potential to charge fees up to $8.9 million per year to recover its costs of issuing permits. Chapter 3 discusses the NPDES permit program's potential for proposing fees.

EPA Reports to OMB Were Not Complete

The EPA's biennial user fee reports to OMB were not complete. The EPA included a summary biennial user fees report, *Biennial User Fees,* in the Agency Financial Report, and submitted more detail in a memorandum to OMB describing the programs with existing fees, proposed fees and fee exceptions. The EPA's summary report in the FY 2011 Agency Financial Report did not discuss the results of the biennial review and any resultant fee proposals. Further, the EPA's more detailed memorandum report to OMB provided incomplete cost information. According to OMB Circular A-25, agencies should discuss the results of the biennial review of user fees and any resultant proposals in the Agency Financial Report. However, the EPA's reports were not complete because the fee programs updated the fee descriptions from the prior biennial review without conducting a thorough cost review, and OCFO did not provide sufficient oversight and validate those reviews. Without complete information, the reports to OMB did not clearly present the condition of EPA fee programs and provide stakeholders with complete information. Transparent processes for reviewing and updating fees help assure payers and others that fees are set fairly and accurately.

We found the following areas where the EPA's reports were not complete:

- The EPA's FY 2011 Agency Financial Report did not discuss the results of the biennial reviews.

- Prior to FY 2010, the EPA reported its financial performance in the annual Performance and Accountability Report but its FY 2009 Performance and Accountability Report did not include a biennial user fee report.

- The EPA's FYs 2008–2009 and 2010–2011 detailed biennial user fee reports included the revenue collected for each existing fee program but not the related program costs. Programs that requested an exception from charging fees did not report their program costs.

- The EPA could not provide documentation to support its transmittal of the FY 2011 detailed biennial user fees memorandum report to OMB.

EPA Did Not Follow OMB Policy to Request User Fee Exceptions

The EPA did not follow the OMB Circular A-25 procedure for requesting user fee exceptions by letter to the OMB Director for programs that did not charge fees. OMB Circular A-25 provides that agency heads request an exception by letter to the OMB Director. EPA program managers and staff did not have the EPA Administrator request fee exceptions by letter to the OMB Director because they were not aware of the OMB requirement to do so. By not getting fee exception approval from OMB, the EPA did not have OMB's assurance that fees were not necessary. The EPA may not have charged appropriate fees in programs to reduce the federal budget deficit.

The OCFO stated it included the programs reporting exceptions in each biennial review detailed report to the OMB. Since the OCFO did not receive feedback from the OMB on its detailed reports, the OCFO believed that its method of reporting agency user fee activities was acceptable to OMB. However, when we asked the OMB about EPA user fee exceptions, the OMB stated that it had not received or granted user fee exceptions to the EPA in recent years. The OMB stated that agency heads may request an exception by letter to the OMB Director.

EPA Did Not Review All Its Programs for Fee Potential

Prior to the FYs 2012–2013 biennial review, the EPA did not review all agency programs to determine whether they should assess fees for government services they provide. OMB Circular A-25 directs the agency to conduct biennial reviews of programs with user charges and all other programs to determine whether it should assess fees for government services it provides. Program office personnel were not aware of the biennial review requirement, and OCFO did not provide internal guidance for reviewing all agency programs. By not reviewing all agency programs, the EPA may not have identified all programs eligible for user fee charges and may have missed an opportunity to charge appropriate fees.

OCFO and program office personnel stated that for biennial reviews conducted through the FYs 2010–2011 review, they reviewed the identified user fee programs but did not review all other agency programs. With our own limited research, we identified the EPA's Underground Injection Control permitting program under the Safe Drinking Water Act as a potential user fee program. The EPA did not include the program in the biennial review, although the program provided a service to specific recipients for which the EPA did not charge fees. This is an example of an activity that the EPA should include in the biennial review to determine whether user fees should be assessed.

Improvements Made for FYs 2012–2013 Biennial Review

Based on the findings of previous OIG audits of fee programs, the OCFO has taken action to help the EPA's fee programs conduct a more thorough FYs 2012–2013 review. The OCFO issued a biennial user fee review guide on March 8, 2013, and conducted biennial user fee review training for the user fee programs on March 21, 2013. The OCFO's guidance for the FYs 2012–2013 review provides the roles and responsibilities, review procedures and summary cost worksheets for completing the review. Some significant provisions are:

- OFM conducts the biennial user fee reviews.
- Program offices provide the costs of the program activities.
- Program offices review all activities for potential new users.
- The Office of Budget collaborates with program offices to review opportunities for new or updated fees.

The OCFO increased its oversight of the FYs 2012–2013 biennial review by assigning an OCFO staff member to each existing fee program to oversee its review. Based on a prior OIG audit recommendation,[3] the OCFO also conducted a FY 2011 biennial cost review of the EPA's Motor Vehicle and Engine Compliance Program in the Office of Air and Radiation. The OCFO used the review experience to help develop its review process for future biennial reviews.

Conclusion

The EPA did not conduct thorough biennial user fee reviews for FYs 2008–2009 and 2010–2011. The EPA should follow the OMB biennial review policy to conduct cost reviews, include complete information in OMB reports, request user fee exceptions, and review all programs for fee potential. By not performing fully compliant biennial user fee reviews in the past, the EPA may not have recovered all related program costs and collected funds that otherwise could have been available to reduce the federal budget deficit. We believe that the EPA could help the federal government reduce the budget deficit by performing more thorough biennial user fee reviews that identify potential additional revenue. We identified an EPA program with the potential to charge fees up to $8.9 million per year to recover its costs of providing a service.

---

[3] EPA OIG report, *EPA Should Update Its Fees Rule to Recover More Motor Vehicle and Engine Compliance Program Costs*, Report No. 11-P-0701, September 23, 2011, recommended that the EPA conduct biennial reviews of the Motor Vehicle and Engine Compliance Program.

Recommendations

We recommend that the Chief Financial Officer:

1. Include in the Agency Financial Report a discussion of the biennial user fee review results and any resultant proposals, to fully comply with OMB Circular A-25 reporting requirements.

2. Coordinate with programs that have claimed an exemption to charging fees under OMB Circular A-25 to have the EPA Administrator request an exception by letter to the OMB Director.

3. Request the fee exception programs to report their program costs to OCFO to provide complete information about program fees and costs and to help determine whether fee exception programs should assess fees.

Preliminary Agency Actions

The OCFO issued a biennial user fee review guide that instructed the program offices to provide costs of the program activities and review all activities for potential users. The OCFO also increased its oversight of the FYs 2012–2013 biennial review. Therefore, we make no recommendations to develop a review guide, conduct cost reviews, review all activities for potential users and increase headquarters oversight.

Agency Comments and OIG Evaluation

The agency agreed with our findings and recommendations and provided intended corrective actions and estimated completion dates.

Chapter 3
NPDES Permit Program
Should Consider Charging Fees

The EPA did not review the NPDES program's permit costs since the early 1990s to determine whether the agency should assess fees for permits in the states, territories and tribes in which the EPA is the permitting authority. Title V of the IOAA authorizes an agency to charge a fee for a service the agency provides. Due to other priorities, the EPA has not considered whether it should assess fees for NPDES permits or reintroduce a proposal for a fee regulation since the EPA withdrew an internal proposal from the early 1990s for a regulation to allow the EPA to charge fees for issuing federal NPDES permits. By not proposing the fees, the EPA may have missed an opportunity to charge appropriate fees to recover its costs. Based on the EPA's recent cost estimate of $8.9 million per year for issuing federal permits, the NPDES program has the potential to charge up to $8.9 million in fees to recover its costs.

EPA Did Not Propose NPDES Fees Since Early 1990s

Although the Clean Water Act does not address the subject of federal fees for federally issued NPDES permits, the IOAA authorizes federal agencies to charge fees for the services they provide. The EPA did not review the NPDES permit program costs since the EPA considered— but withdrew—a proposal from the early 1990s for a regulation under the authority of the IOAA to allow the EPA to charge fees for issuing federal NPDES permits. Due to other priorities, the Office of Water did not consider whether it should assess fees for NPDES permits or attempt a new proposed regulation to charge NPDES fees.

The Office of Water included the NPDES permit program on the FYs 2008–2009 and 2010–2011 user fee exceptions list that it forwarded to the OCFO for inclusion in the biennial user fee reports to OMB. The Office of Water believed that OMB approved an exception for NPDES because the Office of Water did not receive a response from OCFO or OMB about the exception. The Office of Water stated that it assumed that OCFO had received a waiver from OMB that granted the NPDES permit program an exception from charging fees. However, OMB stated that it had not received or approved any fees exception requests from the EPA in recent years. By not proposing NPDES fees, the EPA may have missed an opportunity to charge appropriate fees to recover its costs.

On May 13, 2013, subsequent to the start of our audit, the NPDES permitting program provided an analysis of its estimated annual direct labor costs totaling $8,875,031. Due to time constraints, the cost analysis did not include other direct costs, such as contracts, supplies and travel. According to the Office of Water, the NPDES permit actions may include those types of costs, but the total amount of such costs would not significantly alter the estimate. The analysis also did not include indirect costs needed for full costing, as required by SFFAS No. 4 and OMB Circular A-25. Based on the EPA's cost estimate for issuing federal permits, the NPDES program has the potential to charge up to $8.9 million per year in fees to recover its costs.

Conclusion

According to the OMB's FY 2014 budget overview, the President is committed to continuing to reduce the federal budget deficit. We believe that the EPA could help the federal government in this endeavor by collecting fees to recover its NPDES permitting costs. The EPA has authority under the IOAA to charge fees for NPDES permits, and the OMB Circular A-25 general policy is that user charges will be instituted through the promulgation of regulations. Therefore, the EPA should consider recovering its NPDES permit program costs by charging fees for NPDES permits. The EPA could collect an estimated $8.9 million in fees per year, which would help reduce the federal budget deficit.

Recommendations

We recommend that the Assistant Administrator for Water:

4. Conduct a cost analysis of all direct and indirect costs to determine the EPA's full cost of issuing NPDES permits.

5. Apply federal user fee policy in determining whether to (a) charge fees for issuing federal NPDES permits in which the EPA is the permitting authority, or (b) request an exception from OMB to charging fees.

Agency Comments and OIG Evaluation

The agency agreed with recommendation 4 and provided its corrective action and completion date. The agency did not agree with part of our original recommendation 5 in the draft report to propose a regulation to allow the EPA to charge fees, as appropriate. The Office of Water stated that it was working with the OCFO to request an exception from an NPDES user fee. We agreed with the agency's proposed alternative action and we revised recommendation 5 accordingly. We removed the proposal for a regulation to charge fees and added an option for requesting an exception to fees, to which the Office of Water stated that it will agree. Although the EPA requested a fees exception on December 6, 2013, we consider recommendation 5 to be open until the OMB approves the request.

Status of Recommendations and Potential Monetary Benefits

		RECOMMENDATIONS				POTENTIAL MONETARY BENEFITS (in $000s)	
Rec. No.	Page No.	Subject	Status[1]	Action Official	Planned Completion Date	Claimed Amount	Agreed-To Amount
1	9	Include in the Agency Financial Report a discussion of the biennial user fee review results and any resultant proposals, to fully comply with OMB Circular A-25 reporting requirements.	C	Chief Financial Officer	12/31/13		
2	9	Coordinate with programs that have claimed an exemption to charging fees under OMB Circular A-25 to have the EPA Administrator request an exception by letter to the OMB Director.	C	Chief Financial Officer	12/31/13		
3	9	Request the fee exception programs to report their program costs to OCFO to provide complete information about program fees and costs and to help determine whether fee exception programs should assess fees.	C	Chief Financial Officer	6/30/13		
4	11	Conduct a cost analysis of all direct and indirect costs to determine the EPA's full cost of issuing NPDES permits.	C	Assistant Administrator for Water	6/30/13		
5	11	Apply federal user fee policy in determining whether to (a) charge fees for issuing federal NPDES permits in which the EPA is the permitting authority, or (b) request an exception from OMB to charging fees.	O	Assistant Administrator for Water		$17,800[4]	

[1] O = Recommendation is open with agreed-to corrective actions pending.
  C = Recommendation is closed with all agreed-to actions completed.
  U = Recommendation is unresolved with resolution efforts in progress.

---

[4] OIG's policy is to base efficiencies from recurring events on the projected monetary benefit for the current and following year. The potential monetary benefit represents the recurring cost savings from a proposed regulation to allow the EPA to charge NPDES permit fees, based on the estimated program costs for the current and following year.

Details on Scope and Methodology

We reviewed the EPA's processes for conducting biennial user fee reviews. To gain an understanding of the processes, we:

- Reviewed the applicable laws, federal policy, standards and guidance, and relevant prior audit reports.
- Reviewed the EPA's biennial review procedures and biennial review reports for FYs 2008–2009 and 2010–2011.
- Interviewed personnel in OCFO and program offices with user fees or royalties, proposed user fees and exceptions to user fees.
- Examined OCFO's March 2013 biennial user fee review guide and training materials.
- Reviewed the EPA's 13 program offices' FY 2012 management integrity assurance letters for reported internal control weaknesses.

We interviewed personnel regarding the following programs and their biennial user fee reviews:

- Pesticides maintenance fee (Federal Insecticide, Fungicide, and Rodenticide Act).
- Pesticides registration service fee (Pesticide Registration Improvement Extension Act).
- Premanufacture Notice program.
- Operating permits program (Clean Air Act, 40 Code of Federal Regulations, Parts 70 and 71).
- Research and development licensing fees and royalties.
- Freedom of Information Act.
- NPDES permitting.
- Acid Rain Allowance Transfer program.
- ENERGY STAR program.
- Resource Conservation and Recovery Act permitting.

We examined OCFO's work plan and supporting documentation for its biennial user fee review of the FY 2011 Motor Vehicle and Engine Compliance Program. We did not verify the accuracy of that program's biennial review cost analysis.

We did not assess the reliability of data in any information systems because their use did not materially affect our findings, conclusions or recommendations. We accessed fee collection information in Compass Financials, the agency's accounting system. We did not review the internal controls over Compass Financials from which we obtained financial data, but relied on the review conducted during the audit of the EPA's FY 2012 financial statements.

Prior Reports Reviewed

We reviewed the prior EPA OIG and GAO reports listed in table A-1. The two EPA OIG reports had findings and recommendations related to fee collections and recovery of program costs. The three GAO reports contained information relevant to our review. We used the information and issues disclosed in the EPA OIG and GAO reports to help identify issues as we conducted our audit.

Table A-1: Prior reports reviewed

Report Title	Report No.	Date
EPA Is Not Recovering All Its Costs of the Lead-Based Paint Fees Program	EPA OIG 13-P-0163	February 20, 2013
2012 Annual Report: Opportunities to Reduce Duplication, Overlap and Fragmentation, Achieve Savings, and Enhance Revenue	GAO-12-342SP	February 28, 2012
EPA Should Update Its Fees Rule to Recover More Motor Vehicle and Engine Compliance Program Costs	EPA OIG 11-P-0701	September 23, 2011
Federal User Fees: A Design Guide	GAO-08-386SP	May 29, 2008
Federal User Fees: Some Agencies Do Not Comply with Review Requirements	GAO/GGD-98-161	June 30, 1998

Source: OIG analysis.

EPA OIG Report No. 13-P-0163 disclosed that an EPA program was not collecting enough fees to recover all the costs of administering its program. The EPA had not conducted a formal cost study to determine its actual program costs, and needed to update its fees rule to reflect the amount of fees necessary for the program to recover its costs. The EPA agreed with the report's recommendations and said it planned to update the fees rule and conduct biennial reviews.

GAO Report No. GAO-12-342SP presented cost savings or revenue enhancement opportunities, including GAO's 2011 survey of federal agency fee reviews. The survey responses indicated that for most fees, agencies (1) had not discussed fee review results in annual reports, and (2) had not reviewed the fees and were inconsistent in their ability to provide fee review documentation.

EPA OIG Report No. 11-P-0701 disclosed that an EPA program was not collecting enough fees to recover all reasonable program costs, based on the EPA's rough cost estimate conducted during our audit. The program had not conducted a formal cost study since 2004 to determine its actual program costs, and had not updated the fees rule to recover more costs. The EPA agreed with the report's recommendations and said it planned to update the fees rule and conduct biennial reviews.

GAO Report No. GAO-08-386SP reported on a study of how user fee design characteristics may influence the effectiveness of user fees. GAO examined how the four key design and implementation characteristics of user fees—how fees are set, collected, used and reviewed—

may affect the economic efficiency, equity, revenue adequacy and administrative burden of cost-based fees. The principles outlined in the design guide present a framework for user fee design.

GAO Report No. GAO/GGD-98-161 was a response to a congressional request to review agencies' adherence to the user fee review and reporting requirements in the CFO Act and OMB Circular A-25. The report disclosed that six of the 24 agencies reviewed all of their reported user fees at least every 2 years as required by OMB Circular A-25 during FYs 1993 through 1997, three reviewed all of their reported fees at least once, 11 reviewed some of their reported fees, and four did not review any of their reported fees during this period. The agencies provided various reasons for not reviewing fees, including insufficient cost data and that some of the fees being set by legislation could not be changed without new legislation.

Agency Response to Draft Report

(Received September 27, 2013)

MEMORANDUM

SUBJECT: Response to Office of Inspector General Draft Report, Project No. OA-FY13-0103 "EPA Did Not Conduct Thorough Biennial User Fee Reviews," dated August 12, 2013

FROM: Maryann Froehlich /s/
Acting hief Financial Officer

Nancy K. Stoner, Acting Assistant Administrator /s/
Office of Water

TO: Richard Eyermann, Acting Assistant Inspector General
Office of Audit

Thank you for the opportunity to respond to the issues and recommendations in the subject audit report. The following is a summary of the agency's overall position, along with our responses on each of the report recommendations. For those report recommendations with which the agency agrees, we have provided high-level intended corrective actions and estimated completion dates.

AGENCY'S OVERALL POSITION

The agency concurs with the OIG's overall recommendations. During FY 2013, the EPA improved its biennial review process by issuing a biennial user fee review; conducting webinars to train personnel on the review process, and increasing oversight of these reviews. As part of our implementation guidance we also evaluated the cost of providing the government services and fees for both existing and potential programs.
The agency will:

- Discuss the results of the FY 2013 biennial user fee review in the Agency's Financial Report.

- Make recommendations, to the Office of Management and Budget Director, for exceptions to charge user fees for the EPA programs that meet the criteria outlined in the OMB Circular A-25, Section 6c.

However, the Office of Water does not concur with the recommendation to propose a regulation to allow the EPA to charge fees.

CONTACT INFORMATION

If you have any questions regarding this response, please contact Stefan Silzer, Director of the Office of Financial Management on (202) 564-5389 or Sheila Frace, Deputy Office Director, Office of Wastewater Management, Office of Water on (202) 564-0748.

Attachment

cc: David Bloom
    Joshua Baylson
    Stefan Silzer
    John O'Connor
    Carol Terris
    Jeanne Conklin
    Meshell Jones-Peeler
    Dale Miller
    Sandy Dickens
    Barbara Freggens
    Janet McCabe
    Betsy Shaw
    Maureen Hingeley
    Jim Jones
    Mike Shapiro
    Andrew Sawyers
    Sheila Frace
    Deborah Nagle
    Brian Frazer
    Louis Eby
    Marilyn Ramos
    Lek Kadeli
    Arthur Elkins
    Charles Sheehan
    Aracely Nunez-Mattocks
    Alan Larsen
    Carolyn Copper
    Patricia Hill
    Patrick F. Sullivan
    Paul Curtis
    Arthur Budelier
    Susan Barvenik
    Wendy Swan
    Sheree James

AGENCY'S RESPONSE TO REPORT RECOMMENDATIONS

No.	Recommendation	High-Level Intended Corrective Action(s)	Estimated Completion by Quarter and Fiscal Year
1	Include in the Agency Financial Report a discussion of the biennial user fee review results and any resultant proposals, to fully comply with OMB Circular A-25 reporting requirements.	The Office of the Chief Financial Officer concurs with the overall recommendation. The EPA will include a discussion of the results of the biennial user fee review in the FY 2013 Agency Financial Report AFR.	1st Quarter FY 2014
2	Coordinate with programs that have claimed an exemption to charging fees under OMB Circular A-25, section 6c (2) or (3) to have the EPA Administrator request an exception by letter to the OMB Director.	The OCFO concurs with the overall recommendation. The agency will submit a request to the Director of the OMB for exceptions to charge user fees for the EPA programs that meet the criteria outlined in Circular A-25, Section 6c.	1st Quarter FY 2014
3	Request the fee exception programs to report their program costs to OCFO to provide complete information about program fees and costs and to help determine whether fee exception programs should assess fees.	The OCFO concurs with the overall recommendation. During the agency's FY 2013 biennial user fee review, the agency requested and received the estimated costs from its existing and potential fee programs.	Completed 3rd Quarter FY 2013
4	Conduct a cost analysis of all direct and indirect costs to determine the EPA's full cost of issuing National Pollutant Discharge Elimination System permits.	The Office of Water concurs with the overall recommendation. The OW conducted the required NPDES cost analysis.	Completed 3rd Quarter FY 2013
5	Apply federal user fee policy in determining whether to charge fees for issuing federal NPDES permits in which the EPA is the permitting authority and propose a regulation to allow the EPA to charge fees, as appropriate.	The OW does not concur with the recommendation to propose a regulation to allow the EPA to charge fees. We are working with OCFO to request an exception from a NPDES user fee.	1st Quarter FY 2014

Distribution

Office of the Administrator
Chief Financial Officer
Assistant Administrator for Water
Agency Follow-Up Coordinator
General Counsel
Associate Administrator for Congressional and Intergovernmental Relations
Associate Administrator for External Affairs and Environmental Education
Deputy Chief Financial Officer
Principal Deputy Assistant Administrator for Water
Director, Office of Financial Management, Office of the Chief Financial Officer
Deputy Director, Office of Financial Management, Office of the Chief Financial Officer
Director, Office of Financial Services, Office of the Chief Financial Officer
Deputy Director, Office of Financial Services, Office of the Chief Financial Officer
Audit Follow-Up Coordinator, Office of the Chief Financial Officer
Audit Follow-Up Coordinator, Office of Water
Audit Follow-Up Coordinator, Office of Financial Management, Office of the
        Chief Financial Officer